YOUR KNOWLEDGE HAS VALUE

Rubel Ahmed, B.M. Jewel Rana, S.F. Ahmed

Analytical solution of the MHD free convective unsteady flow over a vertical plate with heat source

GRIN Publishing

Bibliographic information published by the German National Library:

The German National Library lists this publication in the National Bibliography;
detailed bibliographic data are available on the Internet at http://dnb.dnb.de .

Imprint:

Copyright © 2015 GRIN Verlag GmbH
Print and binding: Books on Demand GmbH, Norderstedt Germany
ISBN: 978-3-656-93219-2

This book at GRIN:

http://www.grin.com/en/e-book/295132/analytical-solution-of-the-mhd-free-con-
vective-unsteady-flow-over-a-vertical

Analytical solution of the MHD free convective unsteady flow over a vertical plate with heat source

Rubel Ahmed[l]
Mathematics Discipline
Khulna University, Bangladesh.

B.M Jewel Rana
Mathematics Discipline
Khulna University, Bangladesh.

S.F.Ahmed
Mathematics Discipline
Khulna University, Bangladesh

Abstract: The unsteady free convection and mass transfer boundary layer flow past an accelerated infinite vertical porous plate by taking into account the viscous dissipation is considered when the plate accelerates in its own plane. The dimensionless momentum, energy and concentration equation in the presence of uniform transverse magnetic field has been solved analytically by perturbation technique. The usual similar transformations are applied to the steady momentum, energy and concentration equations and we obtained a set of ordinary differential equations. Then the solutions of the problem of the ordinary differential equations are obtained by using perturbation technique. The expression for velocity field, temperature field, concentration field, skin friction, Nusselt number (Nu) and Sherwood number (Sh) has been found. The results are discussed in detailed with the help of graphs to observe the effect of different parameters.

Keywords: Free convection, mass transfer, unsteady, perturbation technique, viscous dissipation.

I. INTRODUCTION

Consider a two dimensional unsteady flow of a laminar, incompressible, viscous, electrically conducting and heat generation fluid past a semi-infinite vertical moving plate embedded in a uniform porous medium and subjected to a uniform transverse magnetic field in the presence of a pressure gradient has been considered with free convection, thermal diffusion and thermal radiation effects taking in to an account. According to the coordinate system the x^*-axis is taken along the porous plate in the upward direction and y^*axis normal to it. The fluid is assumed to be gray, absorbing–emitting but not scattering medium. Now to solve the momentum, energy and concentration equations usual similarity transformations are introduced. We get a set of ordinary differential equation to obtain the solutions of the problem. The ordinary differential equations are solved by using perturbation technique. The expressions for velocity field, temperature distribution, concentration field, skin friction, Nusselt number (*Nu*) and Sherwood number (*Sh*) have been obtained. The results are discussed in detailed with the help of graphs to analyze the effect of different flow parameters. .

Several workers have studied the problem of free convection flow with mass transfer. Gupta et al [1] have studied heat and mass transfer on a stretching sheet with suction or blowing. Free convection and mass transfer flow through porous medium bounded by an infinite vertical limiting Surface with constant suction have been analyzed by Raptis et al [2]. The free convection and mass transfer flow through a porous medium past an infinite vertical porous plate with time dependent temperature and concentration medium have been discussed by Sattar [3]. Das et al [4] have studied numerical solution of mass transfer effects on unsteady flow past an accelerated vertical porous plate with suction. Viscous dissipation in external natural convection flows have been discussed by

Gebhart B. [5]. Viscous dissipation effects on unsteady free convection and mass transfer flow past an accelerated vertical porous plate with suction have been discussed by Bala Siddulu Malga, Naikoti Kishan. [6].

Chandran et al. [7] have discussed the unsteady free convection flow with heat flux and accelerated motion. Soundalgekar et al. [8] have analyzed the transient free convection flow of a viscous dissipative fluid past a semi-infinite vertical plate.

Free-convection flow with thermal radiation and mass transfer past a moving vertical porous plate have analyzed by Makinde, O. D [9]. Unsteady MHD free convection flow of a compressible fluid past a moving vertical plate in the presence of radioactive heat transfer have been discussed by Mbeledogu, I. U, Amakiri, A.R.C and Ogulu, A, [10]. Numerical Study on MHD free convection and mass transfer flow past a vertical flat plate has been discussed by S. F. Ahmmed [11].

Recently, Das et al [4] have studied numerical solution of mass transfer effects on unsteady flow past an accelerated vertical porous plate with suction. The present study is extension of work; here we considered the effects of viscous dissipation on unsteady free convection and mass transfer boundary layer flow past an accelerated infinite vertical porous flat plate .In their paper they converted the governing equations which are in partial differential equations to ordinary differential equations by introducing similarity variables and then solved the governing equations by finite difference scheme. In the study we have solved the governing partial differential equations only by using the perturbation technique. The effects of the flow parameters on the velocity, temperature and the concentration distribution of the flow field have been studied with the help of graphs. This type of problem has some significant relevance to geophysical and astrophysical studies.

In our present work, we have studied about analytical study on unsteady free convection and mass transfer flow past an accelerated vertical porous plate. The governing equations for the unsteady case are also studied. Then these governing equations are transformed into dimensionless momentum, energy and concentration equations.. The obtained results of this problem have been discussed for the different values of well-known parameters with different time steps. The Wolfram mathematica students for 7 is used to draw graph of the flow.

I.   The governing equation

Introducing a Cartesian co-ordinate system x^* is chosen along the plate in the direction of flow and y^*- axis normal to it. The fluid is assumed to be gray, absorbing–emitting but not scattering medium. Now to solve the momentum, energy and concentration equations usual similarity transformations are introduced.

Within the frame work of delete such assumptions the equations of continuity, momentum, energy and concentration are follows,

$$\frac{\partial v^*}{\partial y^*} = 0 \tag{1}$$

$$\rho\left(\frac{\partial u^*}{\partial t^*} + v^*\frac{\partial u^*}{\partial y^*}\right) = \frac{\partial p^*}{\partial x^*} + \mu\frac{\partial^2 u^*}{\partial y^{*2}} - \rho\beta - \frac{\mu}{K^*}u^* - \sigma B_0^2 u^* \tag{2}$$

$$\frac{\partial T^*}{\partial t^*} + v^* \frac{\partial T^*}{\partial y^*} = \frac{k}{\rho C_p} \frac{\partial^2 T^*}{\partial y^{*2}} - \frac{1}{\rho C_p}\left(\frac{\partial q_r^*}{\partial y^*}\right) - \frac{Q_0}{\rho C_p}(T^* - T_\infty^*) \qquad (3)$$

$$\frac{\partial C^*}{\partial t^*} + v^* \frac{\partial C^*}{\partial y^*} = D_M \frac{\partial^2 C^*}{\partial y^{*2}} + D_T \frac{\partial^2 C^*}{\partial y^{*2}} \qquad (4)$$

where x^*, y^* and t^* are the dimensional distances along the plate, perpendicular to the plate and dimensional time, respectively. u^* and v^* are the components of dimensional velocities along x^* and y^* directions, ρ is the fluid density, μ is the velocity, C_p the specific heat at constant pressure, σ is the fluid electrical conductivity, B_0 is the magnetic induction, K^* is the permeability of the of the porous medium, T^* is the dimensional temperature, D_M is the coefficient of chemical molecular diffusivity, D_T is the coefficient of thermal diffusivity, C^* is the dimensional concentration is the thermal conductivity of the fluid, g is the acceleration due to gravity and q_r^* and R are the local radioactive heat flux and the reaction rate constant respectively.

The boundary conditions for the velocity, temperature and concentration fields are given as follows

where T_w^* and C_w^* are the wall dimensional temperature and concentration respectively. C_∞^* is the free stream dimensional concentration. U_0 and n^* are constants.

From the equation (1), we consider the velocity as the exponential form

$$v^* = -v_0(1 + \varepsilon A e^{n^* t^*}) \qquad (5)$$

where, A is the real positive constant, ε and $A\varepsilon$ are small less than unity and v_0 is a scale of suction velocity which has non-zero positive constant.

In the free stream, from equation (2) we get

$$\rho \frac{dU_\infty^*}{dt^*} = \frac{\partial p^*}{\partial x^*} - \rho_\infty g - \frac{\mu}{K^*} U_\infty^* - \sigma B_0^2 U_\infty^* \qquad (6)$$

Eliminate $\dfrac{\partial p^*}{\partial x^*}$ using equation (2) and equations (6), we obtain

$$\rho\left(\frac{\partial u^*}{\partial t^*}+v^*\frac{\partial u^*}{\partial y^*}\right)=(\rho_\infty-\rho)g+\rho\frac{d\,U_\infty^*}{dt^*}+\mu\frac{\partial^2 u^*}{\partial y^{*2}}-$$
$$\frac{\mu}{K^*}(U_\infty^*-u^*)-\sigma B_0^{\,2}(U_\infty^*-u^*)$$

(7)

$$(\rho_\infty-\rho)=\beta\,(T^*-T_\infty^*)+\beta^*\,(C^*-C_\infty^*)$$

(8)

Substituting equation (4.2.10) into equation (4.2.9), we have

$$\frac{\partial u^*}{\partial t^*}+v^*\frac{\partial u^*}{\partial y^*}=\frac{d\,U_\infty^*}{dt^*}+\mu\frac{\partial^2 u^*}{\partial y^{*2}}+g\beta(T^*-T_\infty^*)+$$
$$g\beta^*(C^*-C_\infty^*)+\frac{\upsilon}{K^*}(U_\infty^*-u^*)+\frac{\sigma B_0^{\,2}}{\rho}(U_\infty^*-u^*)$$

(9)

where, $\upsilon=\dfrac{\mu}{k}$ is the coefficient of the kinematic viscosity.

The radioactive heat flux term by using the Roseland approximation is given by

$$q_r^*=\frac{4\sigma^*}{3k_1^*}\frac{\partial T^{*4}}{\partial y^*}$$

(10)

where, σ^* and k_1^* are respectively the Stefan-Boltzmann constant and the mean absorption coefficient. We assume that the temperature difference within the flow are sufficiently small such that T^{*4} may be expressed as a linear function of the temperature. This is accomplished by expanding in a Taylor series about T_∞^* and neglecting higher order terms, thus

$$T^{*4}\cong 4T_\infty^{*3}-3T_\infty^{*4}$$

(11)

By using equations (10) and (11), into equation (3) is reduced to

$$\frac{\partial T^*}{\partial t^*}+v^*\frac{\partial T^*}{\partial y^*}=\frac{k}{\rho C_p}\frac{\partial^2 T^*}{\partial y^{*2}}-\frac{16\sigma^*T_\infty^{*3}}{3\rho C_p k_1^*}\frac{\partial^2 T^*}{\partial y^{*2}}-\frac{Q_0}{\rho C_p}(T^*-T_\infty^*)$$

(12)

Introducing the non-dimensional quantities and parameters

$$u^*=uU_0\,,v^*=vV_0\,,T^*=T_\infty^*+\theta(T_w^*-T_\infty^*),C^*=C_\infty^*+C(C_w^*-C_\infty^*),U_\infty^*$$

$$u_p^* = U_p U_0, \ K^* = \frac{K v^2}{V_0^2}, \ y^* = \frac{y v}{V_0}$$

$$Gc = \frac{v g \beta^* (C_w^* - C_\infty^*)}{V_0^2 U_0}, Gr = \frac{v g \beta (T_w^* - T_\infty^*)}{V_0^2 U_0}$$

$$Pr = \frac{v \rho C_p}{k}, M = \frac{\sigma B_0^2 v}{\rho V_0^2}, Q = \frac{Q_0 v}{\rho V_0^2 C_p}, R = \frac{4 \sigma^* T_\infty^{*3} (T_w^* - T_\infty^*)}{k_1^* k}, Sc = \frac{v}{D_M}$$

$$t^* = \frac{t v}{V_0^2}, n^* = \frac{V_0^2}{v}$$

Therefore the governing equations in the dimensionless form become equation (13) to equation (16) with the boundary condition (17),

$$\frac{\partial v}{\partial y} = 0 \tag{13}$$

$$\frac{\partial u}{\partial t} + v \frac{\partial u}{\partial y} = \frac{dU_\infty}{dt} + \frac{\partial^2 u}{\partial y^2} + Gr \theta + GmC + N (U_\infty - u) \tag{14}$$

$$\frac{\partial \theta}{\partial t} + v \frac{\partial \theta}{\partial y} = \frac{1}{Pr} \left(1 + \frac{4R}{3} \right) \frac{\partial^2 \theta}{\partial y^2} - Q\theta \tag{15}$$

$$\frac{\partial C}{\partial t} + v \frac{\partial C}{\partial y} = \frac{1}{Sc} \frac{\partial^2 C}{\partial y^2} + S_0 \frac{\partial^2 \theta}{\partial y^2} \tag{16}$$

The corresponding initial and boundary conditions are

$$u = Up, \ \theta = 1 + \varepsilon e^{nt}, \ C = 1 + \varepsilon e^{nt} \qquad \text{at} \quad y = 0$$
$$u \to U_\infty \to 1 + \varepsilon e^{nt}, \ \theta \to 0, C \to 0 \qquad \text{as} \quad y \to \infty \tag{17}$$

II. Solution of the problem

For perturbation technique to solve equations (13) to equation (16) we consider the following series

$$u = u_0(y) + \varepsilon e^{nt} u_1(y) + 0(\varepsilon^2)$$
$$\theta = \theta_0(y) + \varepsilon e^{nt} \theta_1(y) + 0(\varepsilon^2) \tag{18}$$
$$C = C_0(y) + \varepsilon e^{nt} C_1(y) + 0(\varepsilon^2)$$

where, u_0 is mean velocity, θ_0 mean temperature, C_0 mean concentration, n is the frequency of the oscillation, ε is a perturbation quantity and t is the time respectively.

Now, The velocity, temperature and concentration distributions in the boundary layer becomes

$$u(y,t) = 1 + J_1 e^{m_2 y} + J_2 e^{m_6 y} + J_3 e^{m_2 y} + J_4 e^{m_{10} y} + \varepsilon e^{nt} (1 + J_6 e^{m_{10} y} + J_7 e^{m_2 y} + J_8 e^{m_6 y} + J_9 e^{m_2 y} + J_{10} e^{m_4 y}$$
$$+ J_{11} e^{m_2 y} + J_{12} e^{m_8 y} + J_{13} e^{m_6 y} + J_{14} e^{m_2 y} + J_{15} e^{m_2 y} + J_{16} e^{m_4 y} + J_{17} e^{m_{12} y})$$

$$\theta(y,t) = e^{m_2 y} + \varepsilon e^{nt} \left(D_1 e^{m_2 y} + D_2 e^{m_4 y} \right)$$

$$C(y,t) = B_1 e^{m_2 y} + B_2 e^{m_6 y} + \varepsilon e^{nt} \left(B_3 e^{m_6 y} + B_4 e^{m_2 y} + B_5 e^{m_8 y} + D_3 e^{m_2 y} + D_4 e^{m_4 y} \right)$$

It is now important to calculate the physical quantities of primary interest, which are the local wall shear stress, the local surface heat and mass flux. Given the velocity field in the boundary layer, we can now calculate the local wall shear stress i.e., skin friction

$$C_f = \left(\frac{\partial u}{\partial y} \right)_{y=0} = J_1 + J_2 + J_3 + J_4 + \varepsilon e^{nt} (J_6 + J_7 + J_8 + J_9 + J_{10} + J_{11} + J_{12} + J_{13} + J_{14}$$
$$+ J_{15} + J_{16} + J_{17})$$

The dimensionless local surface heat flux i.e., Nusselt number is obtained by

$$Nu = -\left(1 + \frac{4R}{3} \right) \left(\frac{\partial \theta}{\partial y} \right)_{y=0}$$
$$= -\left(1 + \frac{4R}{3} \right) \left[m_2 + \varepsilon e^{nt} \left(m_2 D_1 + m_4 D_2 \right) \right]$$

The Sherwood number is given by

$$Sh = \left(\frac{\partial C}{\partial y} \right)_{y=0}$$
$$= m_2 B_1 + m_6 B_2 + \varepsilon e^{nt} \left(m_6 B_3 + m_2 B_4 + m_8 B_5 \right)$$

III.   Results and discussion

In order to get a clear insight of the physical problem, numerical results are displayed with the help of graphs. This enables us to carry out the numerical calculations for the distribution of the velocity, temperature, concentration, Nusselt number and Sherwood number across the boundary layer for various values of the parameters. To be realistic, the value of Schmidt number (Sc) are chosen for

Sc=0.65. The value of prandtl number (Pr) are chosen for Helium (Pr=0.71, 366K temperature Helium), water (Pr=7.02 at 40^{0}C). Grashof number for heat transfer is chosen to be Gr=1.0. Modified Grashof number for mass transfer is chosen to be Gc=1.0, 2.0. The velocity profiles U for different values of the above parameter are illustrated in Figure 1 to Figure 8, The temperature profiles for different values of the above parameter are displayed in Figure 9 to Figure 11 and the concentration profiles for different values of the above parameter are showed in Figure 12 to Figure 16 .Nusselt number for different values of the above parameter are displayed in Figure 17. Skin friction for different values of the above parameter is let on in Figure 18 Sherwood number for different values of the above parameter is illustrated in Figure 19.

The velocity profile for different values of Grashof number (Gr) are described in Figure 1. From the figure it is observed that an increasing in Gr leads to increase in the values of velocity. Here the Grashof number leads free convection currents. If Gr=0 then it represent the absence of free convection currents. Gr>0 means heating of the fluid of cooling of the boundary surface and Gr<0 means cooling of the fluid of heating of the boundary surface. From the Figure 1 the velocity is highest at y=1.0. The velocity profile for different values of modified Grashof number for mass transfer (Gc) are described in Figure 2. It is observed that an increasing in (Gc) leads to increase in the values of velocity. The curves evince that the peak value of velocity increases rapidly near the wall of the porous plate as modified Grashof number for mass transfer (Gc) increases. At y=1.2 the velocity is highest for Gc=4.0.

Figure 3 evince the velocity for different values of the permeability (K) .it is clear that the peak value of the velocity tends to increases as permeability (K) increases. At y=1.25(approximately) the velocity is highest. For different values of the magnetic field parameters (M), the velocity profiles are plotted in Figure 4. It is obvious that the effect of increasing values of M parameters results in decreasing velocity distribution across the boundary layer. At y=1.0(approximately) the velocity is highest in Figure 4 for magnetic field parameters (M).

Figure 5 exhibit the velocity profiles across the boundary layer for different values of Prandtl number (Pr). It is obvious that the effect of increasing values of Prandtl number (Pr) results in decreasing velocity across the boundary layer. From this figure we see that the velocity profiles increase with an increasing of S_0 from which we conclude that the fluid velocity rises due to greater thermal diffusion. For different values of the Schmidt number (Sc), the velocity profiles are plotted in Figure 7. It is obvious that the effect of increasing values of Sc parameters results in decreasing velocity distribution across the boundary layer. Figure 8 evinces the velocity profiles for different values of the heat source parameters (Q). From the Figure 8 we see that the heat is generated the buoyancy force increases which induce the flow rate to increase giving rise to the increase in the velocity profiles.

The effect of radiation parameter (R) on the temperature profiles is shown in Figure 9. From this figure we observe that the temperature profiles decreases with an increasing of R, with an increasing in the thermal boundary layer thickness. Figure 10 let on the temperature profiles for different values of the heat source parameters (Q). From the Figure 10 we see that the temperature profiles decreases with an increasing of heat source parameters (Q). Figure 11 brings out the temperature profiles across the boundary layer for different values of Prandtl number (Pr). It is obvious that the effect of increasing values of Prandtl number (Pr) results in decreasing temperature across the thermal boundary layer.

The effect of heat source parameters (Q) on the concentrations profiles is shown in Figure 12. From the figure we see that the heat is generated the buoyancy force increases which induces the flow rate to increase giving rise to the increase in the concentrations profiles. Figure 13 let on the concentrations profiles across the boundary layer for different values of radiation parameter (R). It is obvious that the effect of increasing values of radiation parameter (R) results in increasing concentrations across the boundary layer and all curves meet the y axis. The curve is too much high at the wall. Figure 14 brings out the concentrations profiles for different values of Prandtl number (Pr). For different values of the Schmidt number (Sc) the concentrations profiles are plotted in Figure 15. It is observed that the effect of increasing values of Sc parameters results in decreasing concentrations distribution across the boundary layer and all curves meet the y axis. The curve is too much high at the wall. The effect of Soret number (S_0) on the concentrations profiles is shown in Figure 16. From this figure we evince that the concentrations profiles increase with an increasing of S_0.

For different values of the radiation parameter (R) the effect on Nusselt number (Nu) and skin friction are plotted in Figure 17 and Figure 18. It is obvious that the effect of increasing values of radiation parameter (R) results in decreasing Nusselt number (Nu) and increasing the skin friction coefficient. The effect of the Soret number (S_0) on the Sherwood number (Sh) shown in Figure19. It is evince that (Sh) increases when Soret number (S_0) increases.

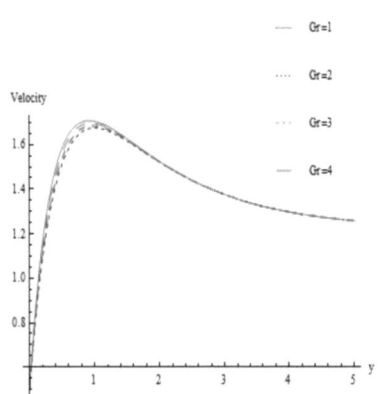

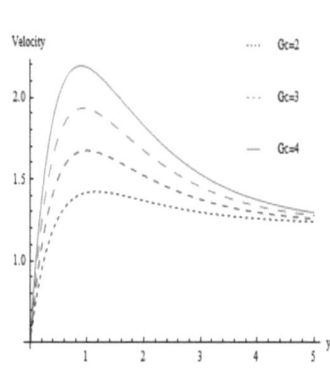

Figure 1: Velocity profiles with Pr=7, R=1 A=0.5 Sc=0.65, M=1, Q=1, ε=0.2, Gc=2, n=0.1 t=1 K=0.5 n=0.1, S_0=1, Up=0.5 for different values of Gr against y.

Figure 2: Velocity profiles with Pr=7 M=1 Sc=0.65 Gr=1, t=1, K=0.5, ε=0.2, R=1 A=0.5 Q=1, n=0.1, S_0=1, Up=0.5 for different values of Gc against y.

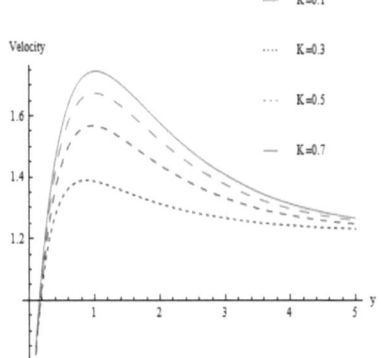

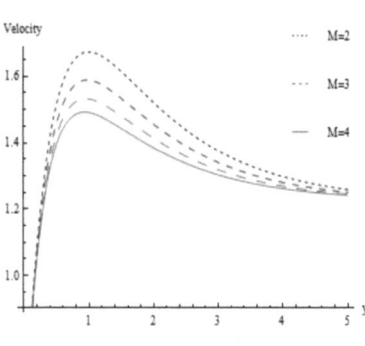

Figure 3: Velocity profiles with M=1 Sc=0.65, Gc=2, Gr=1, t=1, Pr=7, Up=0.5, S_0=1 n=0.1, Q=1, A=0.5, ε=0.2 and R=1 for different values of k against y.

Figure 4: Velocity profiles with Sc=0.65 Gc=2, Gr=1, t=1, K=0.5, Pr=7, n=0.1, Q=1 A=0.5 ε=0.2, R=1, Up=0.5, S_0 =1 for different values of k against y.

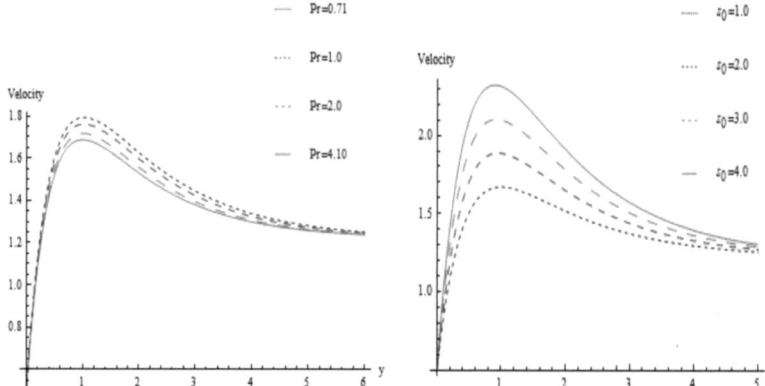

Figure 5: Velocity profiles with M=0.5 Sc=0.65 Gc=2, Gr=1, t=.01, K=0.5, R=1, A=0.5 Q=.01, n=0.1, S_0=1, Up=0.5, ε=0.2 for different values of Pr against y.

Figure 6: Velocity profiles with M=1 Sc=0.65 Gc=2, Gr=1, t=1, K=0.5, Pr=7, n=0.1 Q=1, A=0.5 ε=0.2, R=1, Up=0.5 for different values of against y.

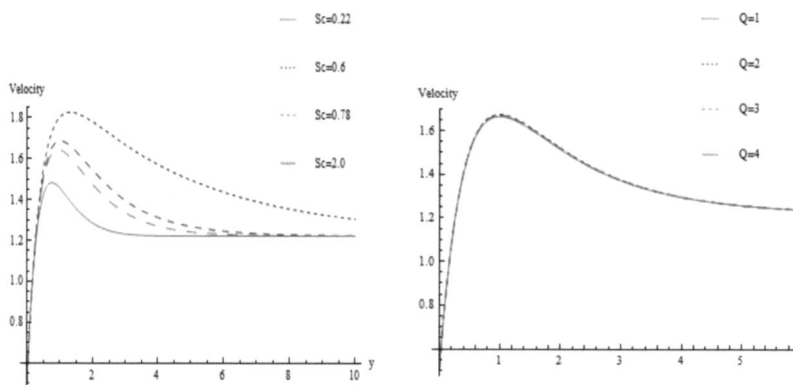

Figure 7: Velocity profiles with M=1 Gc=2 Gr=1, t=1, K=0.5, Pr=7, n=0.1, Q=1, A=0.5 ε=0.2 R=1, Up=0.5 and S_0 =1 for different values of Sc against y.

Figure 8: Velocity profiles with Pr=7, M=1 Sc=0.65 Gc=2, Gr=1, t=1, ε=0.2, K=0.2, R=1 A=0.5 n=0.1 Up=0.5, and S_0 =1 for different values of Q against y.

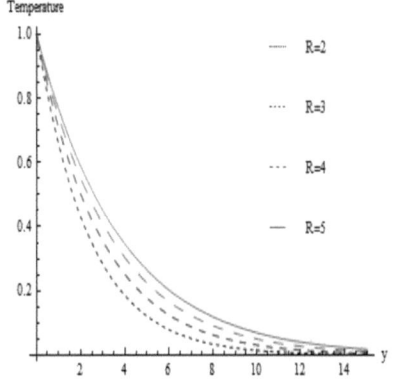

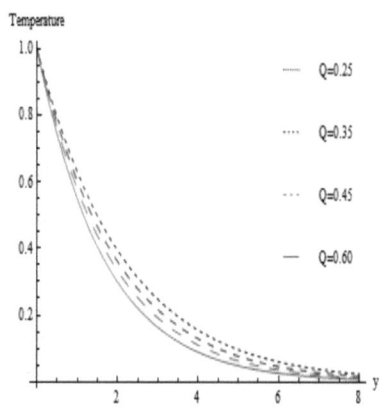

Figure 9: Temperature profile with Pr=7 M=1, Sc=0.65, Gc=2, Gr=1, t=1, K=0.2, A=0.5 n=0.1, Up=0.5, ε=0.02 Pr=0.71, M=1 for different values of R against y.

Figure 10: Temperature profile with Pr=0.71 M=1, Sc=0.65, Gc=2, Gr=1, t=1, K=0.5 ε =0.02 R=1, A=0.5 n=1, Up=0.5 for different values of Q against y.

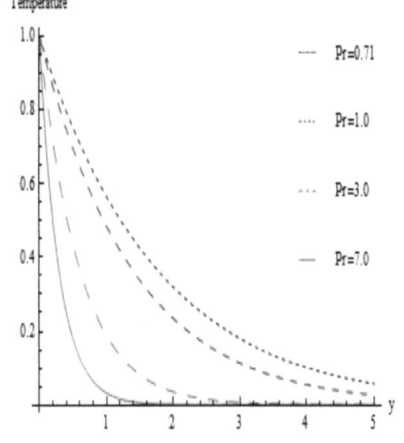

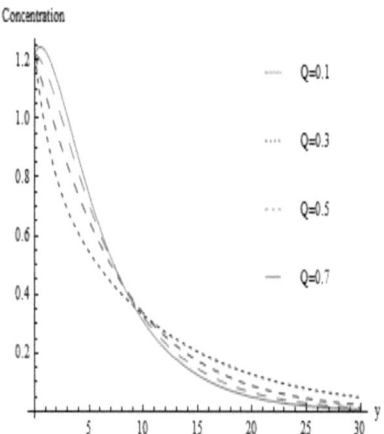

Figure 11: Temperature profile with Pr=0.71 M=1 Sc=0.65, Gc=2, Gr=1, t=1, K=0.5, ε =0.02, R=1 A=0.5 n=1,S_0 =1, Up=0.5 for different values of Pr against y.

Figure 12: Concentration profile with Gc=2 Pr=0.71, M=1, Sc=0.65, Gr=1, t=1, K=0.5 ε =0.02 R=1, A=0.5 n=1, S_0=1, Up=0.5 for different values of Q against y.

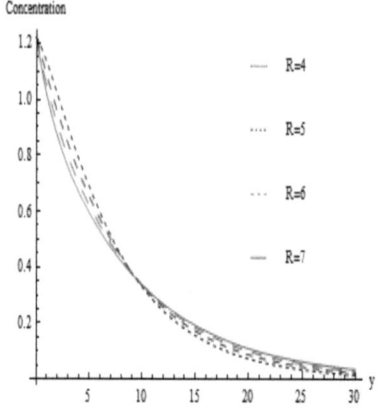

Figure 13: Concentration profiles with Gc=2, Pr=0.71, M=1, Sc=0.65, Gr=1, t=1 K=0.5, ε=0.02, A=0.5, Q=0.5, n=1, S₀=1 Up=0.5 for different values of R agains y.

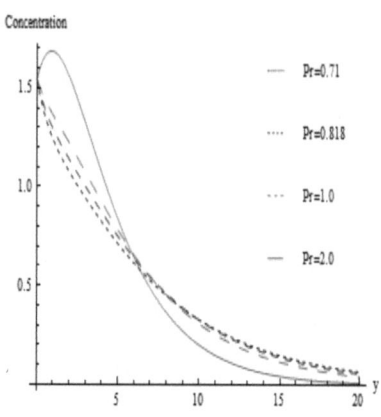

Figure 14: Concentration profiles with Gc=2 M=1, Sc=0.65, Gr=1, t=1, K=0.5, ε =0.02 A=0.5 Q=0.5, n=1, S₀=1, Up=0.5, R=3.5 for different values of Pr against y.

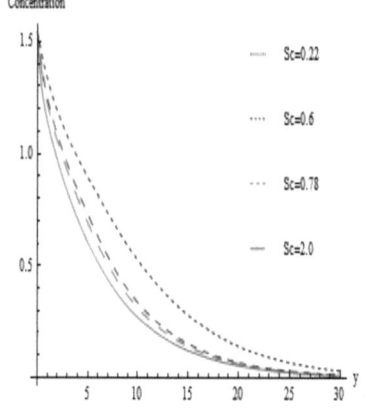

Figure 15: Concentration profiles with Gc=2, M=1, Sc=0.65, Gr=1, t=1, K=0.5, ε =0.2 A=0.6, Q=0.5, n=0.1, S₀ =1, Up=0.5, R=0.1 and Pr=0.71 for different values of Sc against y.

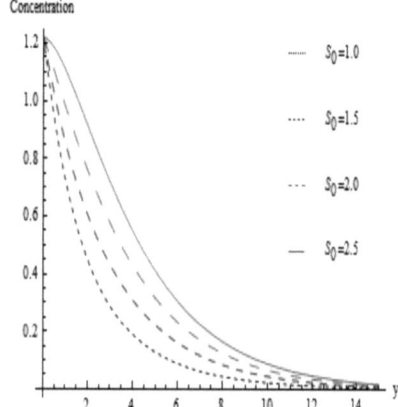

Figure 16: Concentration profiles with Gc=2 M=1, Sc=0.65, Gr=1, t=1, K=0.5, ε =0.02, A=0.5 Q=0.5, n=1, S₀=1, Up=0.5, R=3.5 for different values of Pr against y.

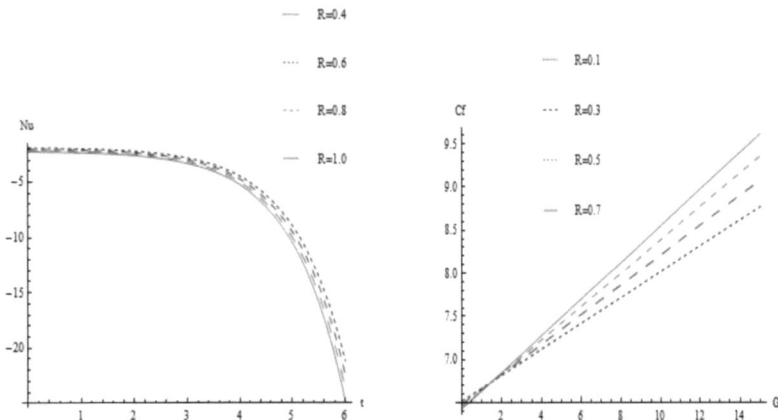

Figure 17: Nusselt number profiles with Sc=0.65, K=0.5, ε =0.02, A=0.5, Q=2.0, R=1.0 and Pr=0.71 for different values of R against t.

Figure 18: Sherwood number profiles Sc=0.65,ε =0.2, A=0.6, Q=0.5, n=0.1, =1, Up=0.5, R=1.0 and Pr=0.71 for different values of versus t.

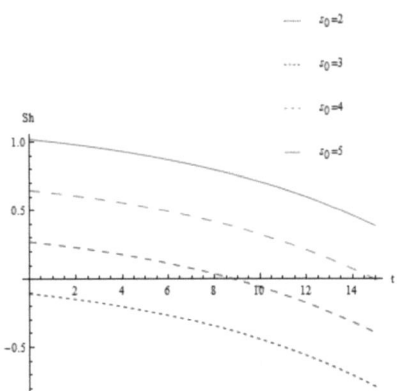

Figure 19: Sherwood number profiles Sc=0.65,ε =0.2, A=0.6, Q=0.5, n=0.1, =1, Up=0.5, R=1.0 and Pr=0.71 for different values of versus t.

IV.   Conclusion

In the present research work, boundary layer equations become non-dimensional by using non-dimensional quantities. The non-dimensional boundary layer equations are nonlinear partial differential equations. These equations are solved perturbation method. Results are given graphically to display the variation of velocity, temperature, concentration, Nusselt number, Skin friction and Sherwood number. The following conclusions are set out through the overall observations.

1) , In presence of magnetic field velocity increases with an increase of Grashof number (Gr), modified Grashof number for mass transfer (Gc), Permeability (k) and Soret number (S_0). Whereas velocity decreases with an increase of Prandtl number (Pr), magnetic field parameter (M) and Scmidth number (Sc). There is no effect of heat source parameter (Q) on the velocity profiles.

2) The temperature increases with an increase of radiation parameter (R) and temperature decreases with the increasing value Prandtl number (Pr), source parameter (Q).

3) The Nusselt number (Nu) increases with an increase of radiation parameter (R).

4) The Skin friction increases with an increase of radiation parameter (R) and the Sherwood number increases with an increase of Soret number (S_0).

V.   References

[1] P.S. Gupta and A.S. Gupta, "Heat and mass transfer on a stretching sheet with suction or blowing". *Can J Chem Eng.*, 55, 744–746, 1977.

[2] A.Raptis. "Free convection and mass transfer flow through porous medium bounded by an infinite vertical limiting surface with constant suction." G.T Zivnidis, N.Kafousis, *Letters in heat and mass transfer,* 8, 5,417-424, 1981.

[3] M. A.Satter, "Free convection and mass transfer flow through a porous medium past an infinite vertical porous plate dependent temperature and concentration." *Int.J. Pure Appl.Math.* 23, 759-766, 1994.

[4] S.S. Das, S.K. Sahoo and G.C. Dash, Bull, "Numerical solution of mass transfer effects on unsteady flow past an accelerated vertical porous plate with suction."*Math. Sci. Soc., 2,* 29,(1), 33–42, 2006.

[5] B.Gebhart andMollendorf, "Viscous dissipation in external natural convection flows." *Journal of Fluid Mechanics,* 38, 97-107, 1969.

[6] Bala Siddulu Malga, Naikoti Kishan. " Viscous dissipation effects on unsteady free convection and mass transfer flow past an accelerated vertical porous plate with suction". *Pelagia research library advances in applied science research*, 2011, 2 (6):460-469.

[7] P.Chandran, N.C.Sacheti and A.K.Singh, "Unsteady free convection flow with heat flux and accelerated motion." *J. Phys. Soc. Japan* 67,124-129, 1998.

[8] V. M. Soundalgekar, B. S. Jaisawal, A. G. Uplekar and H. S. Takhar, "The transient free convection flow of a viscous dissipative fluid past a semi-infinite vertical plate." *J. Appl. Mech. Engng.4*, 203-218, 1999.

[9] O. D Makinde "Free-convection flow with thermal radiation and mass transfer past a moving vertical porous plate*", Int. Comm. Heat Mass Transfer, 32, pp.* 1411– 1419, 2005

[10] Mbeledogu, I. U, Amakiri, A.R.C and Ogulu "Unsteady MHD free convection flow of a compressible fluid past a moving vertical plate in the presence of radioactive heat transfer", *Int. J. of Heat and Mass Transfer, 50, pp.* 1668– 1674, 2007.

[11] S. F. Ahmmed, S. Mondal and A. Ray." Numerical studies on MHD free convection and mass transfer flow past a vertical flat plate". *IOSR Journal of Engineering (IOSRJEN). ISSN* 2250-3021, vol. 3, Issue 5, pp 41-47, May. 2013.

VI. Appendix

$$m_2 = -\frac{1+\sqrt{1+4Q\beta_1}}{2\beta_1}$$

$$m_4 = -\frac{1+\sqrt{1+4(n+Q)\beta_1}}{2\beta_1}$$

$$m_6 = -Sc$$

$$m_8 = -\frac{Sc+\sqrt{(Sc)^2+4nSc}}{2}$$

$$m_{10} = -\frac{1+\sqrt{1+4N}}{2}$$

$$m_{12} = -\frac{1+\sqrt{1+4(N+n)}}{2}$$

$$\beta_1 = \frac{(3+4R)}{3Pr}$$

$$D_1 = \frac{-Am_2}{\beta_1 m_2{}^2 + m_2 - (n+Q)}$$

$$D_2 = (1-D_1)$$

$$D_3 = \frac{-ScS_0 m_2{}^2 D_1}{m_2{}^2 + Scm_2 - nSc}$$

$$D_4 = \frac{-ScS_0 m_4{}^2 D_2}{m_4{}^2 + Scm_4 - nSc}$$

$$B_1 = \frac{-ScS_0 m_2}{m_2 + Sc}$$

$$B_2 = (1-B_1)$$

$$B_2 = \frac{-AScm_6 B_2}{m_6{}^2 + Scm_6 - nSc}$$

$$B_4 = \frac{-AScm_2 B_1}{m_2{}^2 + Scm_2 - nSc}$$

$$B_5 = (1-B_3-B_4-D_3-D_4)$$

$$J_1 = \frac{-Gr}{m_2{}^2 + m_2 - N}$$

$$J_2 = \frac{-GcB_2}{m_6{}^2 + m_6 - N}$$

$$J_3 = \frac{-GcB_1}{m_2{}^2 + m_2 - N}$$

$$J_4 = (U_p - 1 - J_1 - J_2 - J_3)$$

$$J_6 = -\frac{AJ_4 m_{10}}{m_{10}{}^2 + m_{10} - (N+n)}$$

$$J_7 = -\frac{AJ_1 m_2}{m_2{}^2 + m_2 - (N+n)}$$

$$J_8 = -\frac{AJ_2 m_6}{m_6{}^2 + m_6 - (N+n)}$$

$$J_9 = -\frac{AJ_3 m_2}{m_2{}^2 + m_2 - (N+n)}$$

$$J_{10} = -\frac{GrD_2}{m_4{}^2 + m_4 - (N+n)}$$

$$J_{11} = -\frac{GrD_1}{m_2{}^2 + m_2 - (N+n)}$$

$$J_{12} = -\frac{GcB_5}{m_4{}^2 + m_4 - (N+n)}$$

$$J_{13} = -\frac{GcB_3}{m_6{}^2 + m_6 - (N+n)}$$

$$J_{14} = -\frac{GcB_4}{D^2 + D - (N+n)}$$

$$J_{15} = -\frac{GcD_3}{m_2{}^2 + m_2 - (N+n)}$$

$$J_{16} = -\frac{GcD_4}{m_4{}^2 + m_4 - (N+n)}$$

$$J_{17} = -(1+J_7+J_8+J_9+J_6+J_{11}+J_{10} + J_{13}+J_{14}+J_{12}+J_{15}+J_{16})$$